Published by

Revolutionary Hearts Industries

Illustrated by Naomi Winston
Created in Partnership with Early Charm

To someone with a future in STEM,

Early Charm works with the world's leading scientists to make new products from their discoveries. We invent and make 3D printed parts for rockets, cameras that see cancer inside of people, sonars used on aquafarms to count shrimp, shirts that measure breathing, uniforms that protect soldiers from mosquitos, software that designs new life saving drugs, and much more.

To do all this we need creative teams. The most creative teams are always filled with the most diverse people. Each sees and understands the world differently, shares those views with their teammates, and respects the views of others. Together, they solve the world's toughest problems.

At Early Charm we embrace diversity. You can see a bit of it in this book. People of different races and religious beliefs, people from India, Bulgaria, China, Korea, Vietnam and Nepal and from seven different states, and people with science, engineering, arts, journalism and business degrees are all represented here.

We hope this book helps you see yourself working in a science, technology, engineering or math company some day.

LETTER FROM EARLY CHARM

YOU BELONG HERE: DIVERSITY IN STEM

What job do you want to have when you grow up?

Write 3-5 sentences explaining the job that you want to have.

The job that I want to have when I grow is:

_____

_____

_____

_____

_____

_____

_____

_____

To help you find belonging,

"What do you want to be when you grow up?" is one of the most common questions that someone will ask you not only as you go through school but even after you graduate. The answer, for most, will change several times and that is okay!

I hope that this book--filled with immigrants, artists, and visionaries--helps you realize that it doesn't matter where you grow up but who you grow up to be. Your destiny is not defined by your circumstances but by how you leverage your circumstances to create purpose in the world around you.

Remember that whatever room you walk in, you deserve to be there. You worked hard. You are smart. You belong. Age, race, gender, and a multitude of other factors can affect how you are able to exist in the world but please don't let that stop you.

Your insight, your vision, and your dreams of the future are needed to create a better society not only for those in your life but those who exist across the world. I hope that the people in this book are able to showcase to you the importance of perseverance, dedication, and an unwavering belief in self.

Remember, that everything has a purpose and you are everything.

LETTER FROM THE AUTHOR

YOU BELONG HERE: DIVERSITY IN STEM

"Find something interesting to you that energizes you. Chase that without worrying about the rest."

Johns Hopkins University| Ph.D.|Materials Science & Engineering

Stephen Farias

Stephen created a company to make better cell phones. He also helps all the other scientists solve problems.

"Craft the future one layer at a time!"

University of Louisville| Ph.D. | Mechanical Engineering

Kavish Sudan

Kavish studies how to make strong parts for electric motors and rockets.

"Science allows you and makes you adept at questioning, which is vital to keep moving forward."

Worcester Polytechnic Institute| Masters| Chemical Engineering

Ronish Shrestha

Ronish makes new materials for bandages that will heal wounds faster.

"Experiment, Fail, Learn, Repeat"

Georgia Tech| B.S. | Chemical Engineering

Kelli Booth

Kelli created Early Charm and helps all the employees make good decisions.

"Never dim your light for someone else's comfort. The world needs your brilliance."

Yonsei University| B.S. | Informational and Industrial Engineering

KC Song

KC writes computer software that tells shrimp farmer when to feed their shrimp.

"Follow your curiosity! Your discoveries may change the world one day."

University of Maryland| Ph.D. |Fisheries Science

Suzan Shahrestani

Suzan invented a sonar to count shrimp on farms so t[...] farmer knows how much to feed them.

“Whether you think you can, or you think you can't -- you're right”

University of California| Ph.D. | Chemical Engineering

Quynh Vo

Quynh writes computer software that is used to discover new drugs to fight diseases.

"Don't be afraid to change direction and try something different, whether it is your major in school, your career path, or a project you are working on."

University of Maryland| Ph.D. | Environmental Molecular Biology

Kelsey Abernathy

Kelsey works with inventors to get their invention manufactured and sold to customers.

"Career paths are not set in stone based on the subject you study. You can choose to contribute through a variety of job roles based on what suits you."

University of Washington| M.S. | Chemistry

Tina Guvench

Tina works with scientists to plan their work so they don't run out of money.

"Don't be sorry. Be Better."

Purdue University Global | Masters Business Administration

Samir Rashiduddin

Samir keeps track of how much money customers pay for products and how much it costs to make products

"Keep that ideal person in your life and work towards your goals day by day."

University of Kentucky| Ph.D. | Chemical & Materials Engineering

Abhishek Kognole

 Abhishek uses computer models of proteins to design drugs to stop cancer.

"Don't ever be afraid of trying new things because that is the only way you are going to grow."

Towson University| Masters | Applied Physics

Itso Ivanov

Itso manages a team of engineers that manufacture products by 3D printing.

"Anything is possible."

MICA | B.A. | Painting

Winston Frazer

Winston created a company to make 3D printed cover for artificial limbs.

"You belong there even if someone believes that you don't."

Coppin University | Masters | Polymer Sciences

Keturah Postell

Keturah protects soldiers by making their uniforms hard to catch on fire.

"Do the work to improve the lives of others."

Idaho State University| Ph.D. | Pharmaceutical Science

Oliver Tao

Oliver works with customers to make sure that software they buy runs on different types of hardware

> "Do the best you can until you know better. Then, when you know better, do better."
> --Maya Angelou

University of Maryland, College Park | B.S.| Journalism

Morgan Eichensehr

Morgan writes about how new science discoveries can help people live longer and be happier.

"Your worth is internally based not externally given."

University of Central Florida| B.S. | Mechanical Engineering

Nathan Nimbargi

Nathan sells computer programs to companies that develop new drugs to fight diseases.

"Keep going one day at a time and one day you'll realize that you're living the dream you have always wanted. KEEP GOING!"

Johns Hopkins Whiting School of Engineering | Masters | Chemical and Bimolecular Engineering

Rashi Sultania

Rashi tests new drugs on cells to make sure they work properly.

(Your Quote Here)

_____

Draw yourself in the job you want to have and write yourself something nice to keep you motivated!

Draw Yourself and a Background for Your Job

What is one college in this book you want to learn more about?

_____

Research 1 of the colleges mentioned in the book and write 3-5 sentences about it.

The college I want to learn more about is:

_____

_____

_____

_____

_____

_____

_____

_____

What is a major from this book that you want to know more about?

_____

Research 1 of the majors (what people studied in college) mentioned in the book and write 3-5 sentences about it.

The college major that I want to know more about is:

_____

_____

_____

_____

_____

_____

_____

_____

_____

Which career from this book would you like to learn more about?

Write 5 sentences saying which career you want to learn more about and why you want to learn more about it.

I want to learn more about
(   ) because

_____

_____

_____

_____

_____

_____

_____

_____

Creative Representation as a Movement for Change

Naomi Winston

Completing this book with the amazing team, leadership, and community around Early Charm truly reminded me of the importance of community centered education. I hope that you enjoyed not only the beautiful images but also the incredible stories that are behind each and every one.

There is no doubt that sometimes striving for social change, trying improve the lives of others, and working to better yourself every single day is hard. I never believed that making coloring books could be my destiny in life but there were so many other people, even if I didn't know them well, that believed it for me. We all need people to believe in us, for us.

This book is filled with so many of those people. People who have triumphed, who have failed, and who have succeed with kindness throughout. You are not alone in this world and truly trust us when we say...you belong!

Everything has a purpose, and you are everything.

This coloring book was brought to you in partnership with

Early Charm welcomes school tours at our facility in the Pigtown neighborhood of Baltimore. Have a teacher or school administrator send a request to **info@earlycharm.com**

 Scan the QR code or visit their
website to learn more!

_____

https://www.earlycharm.com/